EARTH MATERIALS AND SYSTEMS

ROCKS

by Tamra B. Orr

PEBBLE
a capstone imprint

Pebble Explore is published by Pebble, an imprint of Capstone.
1710 Roe Crest Drive
North Mankato, Minnesota 56003
www.capstonepub.com

Library of Congress Cataloging-in-Publication Data
Names: Orr, Tamra, author.
Title: Rocks / Tamra B. Orr.
Description: North Mankato, MN : Capstone Press, [2021] | Series: Earth materials and systems | Includes index. | Audience: Grades 4-6
Identifiers: LCCN 2020000834 (print) | LCCN 2020000835 (ebook) | ISBN 9781977123800 (hardcover) | ISBN 9781977126801 (paperback) | ISBN 9781977124173 (pdf)
Subjects: LCSH: Rocks—Juvenile literature.
Classification: LCC QE432.2 .O77 2021 (print) | LCC QE432.2 (ebook) | DDC 552—dc23
LC record available at https://lccn.loc.gov/2020000834
LC ebook record available at https://lccn.loc.gov/2020000835

Image Credits
iStockphoto: abadonian, 10, Francois Boudrias, 9, goncharovaia, 25, Ilia Bystrov, 14, JacobH, 13, noblige, 5, Siobhan_Fraser, 28, slavemotion, 17, t-lorien, 18; Shutterstock Images: Clara Bastian, 21, Hecos, 27, Randy McVeigh, 26, Red Tiger, 23, robert cicchetti, 6, superoke, cover, Yuriy Buyvol, 7
Design Elements: Shutterstock Images

Editorial Credits
Editor: Charly Haley; Designer: Jake Nordby; Production Specialist: Joshua Olson

All internet sites appearing in back matter were available and accurate when this book was sent to press.

Printed in the United States 6537

TABLE OF CONTENTS

Words in **bold** are in the glossary.

What Are Rocks?

Rocks are all around us. You can see them outdoors. Small pebbles on the sidewalk are rocks. Large boulders on mountains are rocks too. Many rocks are underground.

Some rocks are smooth. Others are rough. Rocks are all shapes and sizes. Wind and water can change a rock's shape over time.

All rocks are made of at least one **mineral**. Most rocks are made of more than one mineral. Minerals are things in nature that are not alive.

A rock's color comes from its minerals. Rocks made from different minerals are different colors.

How Do Rocks Form?

There are three types of rocks. One type is called igneous rock. Igneous rocks are made from melted rock that has cooled. Melted rock is deep underground. It is called **magma**. It is very hot.

This melted rock comes out of **volcanoes**. It becomes **lava**. The lava flows across the ground. The lava cools as it moves. It becomes hard rock. This is called igneous rock.

lava

Igneous rocks

Some igneous rocks are smooth and shiny. They are made when lava cools quickly. Other igneous rocks are rough. They are made when lava cools slowly.

Igneous rocks also form when magma cools underground. These rocks are called **granite**.

A second type of rock is called sedimentary rock. These are rocks made from small pieces such as sand or pebbles.

Sedimentary rocks are made when these pieces get buried. The small pieces press together. They become hard rock. Then the rock gets buried under more small pieces. The new pieces harden. They form a new **layer**.

The rock keeps getting buried. More layers form. The rock gets bigger. This happens over millions of years.

Dead plants and animals can help form sedimentary rocks too. Dead animals and plants break down. They become part of the rocks.

layers

The third type of rock is called metamorphic rock. These rocks are made from other rocks.

Rocks change when they get very hot. This happens to rocks that are buried deep within Earth. Buried rocks also change when the ground pushes down on them. These changes make new minerals in the rocks. The rocks become metamorphic rocks. These changes happen over millions of years.

Where Are Rocks?

Rocks are everywhere. Most rocks that people find outdoors are sedimentary rocks. These rocks are often found near water. Moving water in rivers or oceans breaks rocks into small pieces. These pieces eventually form sedimentary rocks.

These rocks are found in deserts too. Wind blows fast across flat desert land. This wind hits rocks. It breaks them into pieces over time. The pieces form sedimentary rocks.

People sometimes use machines to get metamorphic rocks out of mountains.

People can also find metamorphic rocks outdoors. These rocks form deep underground. But they can get pushed above ground over time.

People usually find these rocks near mountains. Mountains are in places where the ground moves a lot. This happens slowly. The moving dirt pushes on buried rocks. This makes metamorphic rocks.

Igneous rocks can be found near volcanoes. But many are in the ground. These rocks make up most of the rocks in Earth's **crust**. Earth has many layers underground. Earth's crust is the top layer. It is the layer we walk on. Most of the ocean floor is also made of these rocks.

volcano

Rocks are always changing. The changes happen very slowly over millions of years. Rocks change into different types of rocks. This is called the rock cycle.

Rocks form deep underground. They heat and cool. They break or bend. They are pushed above ground. Rain and wind move them. Some break into small pieces. The pieces get buried and harden. All of these things change rocks. Any rock can change into a different kind of rock.

Why Are Rocks Important?

We use rocks for many different things. Some rocks are used to build buildings. Bricks are made of rocks. The concrete in sidewalks is made of rocks too.

Many things that we use every day are made of rocks. Rocks are used to make some of the dishes we use. An igneous rock called pumice is used to make blue jeans. This rock is rubbed on fabric. It makes the fabric look like the jeans we wear.

Rocks make homes for many plants and animals. Some things live only on mountains. They would not have homes without these rocky places.

Rocks also help feed plants. The minerals in rocks get into soil. Minerals help plants grow.

2X

Rocks can help us learn about the past. Scientists study rocks. They learn about what Earth was like long ago. Rocks can show scientists if a place once had a volcano. Rocks can show if a place was once underwater.

We see rocks around us every day. Rocks are always slowly changing. They come in many forms, and many living things use them. Rocks help us learn about Earth.

Glossary

crust (KRUHST)—the top layer of Earth

granite (GRAN-it)—a hard igneous rock used to build buildings

lava (LAH-vuh)—hot liquid rock that comes out of a volcano

layer (LAY-ur)—a thickness of something laid on top of another

magma (MAG-muh)—melted rock found beneath the Earth's surface; magma flows as lava out of volcanoes and becomes igneous rock when it cools

mineral (MIN-ur-uhl)—a solid substance in the ground made by nature that is not a plant or animal; minerals are found in rocks and soil

volcano (vol-KAY-noh)—a mountain that can erupt to let out lava

Read More

Dennie, Devin. *My Book of Rocks and Minerals.* New York: DK Penguin Random House, 2017.

Oxlade, Chris. *Rocks.* Chicago: Heinemann Raintree, 2016.

Sipperley, Keli. *Fossils.* North Mankato, MN: Capstone Press, 2021.

Internet Sites

DK Find Out!: Rocks and Minerals
https://www.dkfindout.com/us/earth/rocks-and-minerals/

National Geographic Kids: Geology 101
https://kids.nationalgeographic.com/explore/science/geology-101/

Wonderopolis: How Do Rocks Get Their Colors?
https://www.wonderopolis.org/wonder/how-do-rocks-get-their-colors

Index